¡Criaturas diminutas!/Bugs, Bugs, Bugs!

Mantis religiosas/Praying Mantises

por/by Margaret Hall

Traducción/Translation: Dr. Martín Luis Guzmán Ferrer
Editor Consultor/Consulting Editor: Dra. Gail Saunders-Smith

Consultor/Consultant: Gary A. Dunn, MS, Director of Education
Young Entomologists' Society Inc.
Lansing, Michigan

Mankato, Minnesota

Pebble Plus is published by Capstone Press,
151 Good Counsel Drive, P.O. Box 669, Mankato, Minnesota 56002.
www.capstonepress.com

Printed in the United States of America

1 2 3 4 5 6 11 10 09 08 07 06

Library of Congress Cataloging-in-Publication Data
Hall, Margaret, 1947–
[Praying mantises. Spanish & English]
Mantis religiosas = Praying mantises/de/by Margaret Hall.
p. cm.—(Pebble plus. ¡Criaturas diminutas!/Bugs, bugs, bugs!)
Includes index.
ISBN-13: 978-0-7368-6681-1 (hardcover)
ISBN-10: 0-7368-6681-7 (hardcover)
1. Praying mantis—Juvenile literature. I. Title. II. Title: Praying mantises. III. Series: Pebble plus. ¡Criaturas diminutas! (Spanish & English)
QL505.9.M35H25 2007
595.7'27—dc22 2005037470

Summary: Simple text and photographs describe the physical characteristics and habits of praying mantises—in both English and Spanish.

Editorial Credits
Sarah L. Schuette, editor; Katy Kudela, bilingual editor; Edia del Risco, Spanish copy editor; Linda Clavel, set designer; Kelly Garvin, photo researcher; Karen Hieb, product planning editor

Photo Credits
Bruce Coleman Inc./J&L Waldman, 12–13; Karen McGougan, 8–9
Minden Pictures/Gerry Ellis, 18–19
Pete Carmichael, 4–5, 6–7, 15, 16–17
Robert & Linda Mitchell, cover, 1, 11
Stephen McDaniel, 20–21

Note to Parents and Teachers

The ¡Criaturas diminutas!/Bugs, Bugs, Bugs! set supports national science standards related to the diversity of life and heredity. This book describes praying mantises in both English and Spanish. The images support early readers in understanding the text. The repetition of words and phrases helps early readers learn new words. This book also introduces early readers to subject-specific vocabulary words, which are defined in the Glossary section. Early readers may need assistance to read some words and to use the Table of Contents, Glossary, Internet Sites, and Index sections of the book.

Table of Contents

Tabla de contenidos

Praying Mantises

What are praying mantises?

Praying mantises are insects.

Las mantis religiosas

¿Qué son las mantis religiosas?

Las mantis religiosas son insectos.

How Praying Mantises Look

Praying mantises are about
the size of a child's finger.
Praying mantises have six legs.

Cómo son las mantis religiosas

Las mantis religiosas son como
del tamaño del dedo de un niño.
Las mantis religiosas tienen seis patas.

Many praying mantises have
green or brown bodies.
Praying mantises can also
be pink, white, or yellow.

Muchas de las mantis religiosas
tienen el cuerpo verde o marrón.
Las mantis religiosas también pueden
ser de color rosa, blancas o amarillas.

Some praying mantises
look like leaves or flowers.
Praying mantises sit
very still.

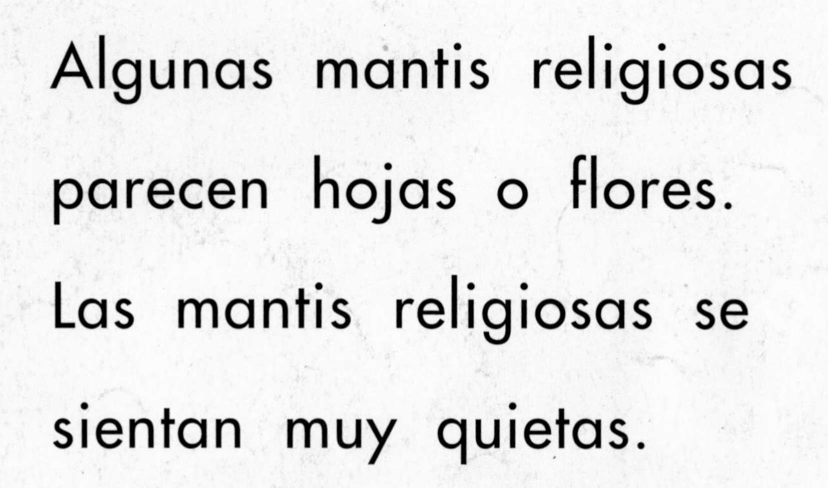

Algunas mantis religiosas
parecen hojas o flores.
Las mantis religiosas se
sientan muy quietas.

What Praying Mantises Do

Praying mantises turn their heads from side to side. Their heads look like triangles.

Qué hacen las mantis religiosas

Las mantis religiosas voltean la cabeza de un lado a otro. Sus cabezas parecen triángulos.

Praying mantises fold
their front legs together.

Las mantis religiosas doblan sus
patas delanteras al mismo tiempo.

Praying mantises grab other animals to eat.

Las mantis religiosas atrapan a otros animales para comérselos.

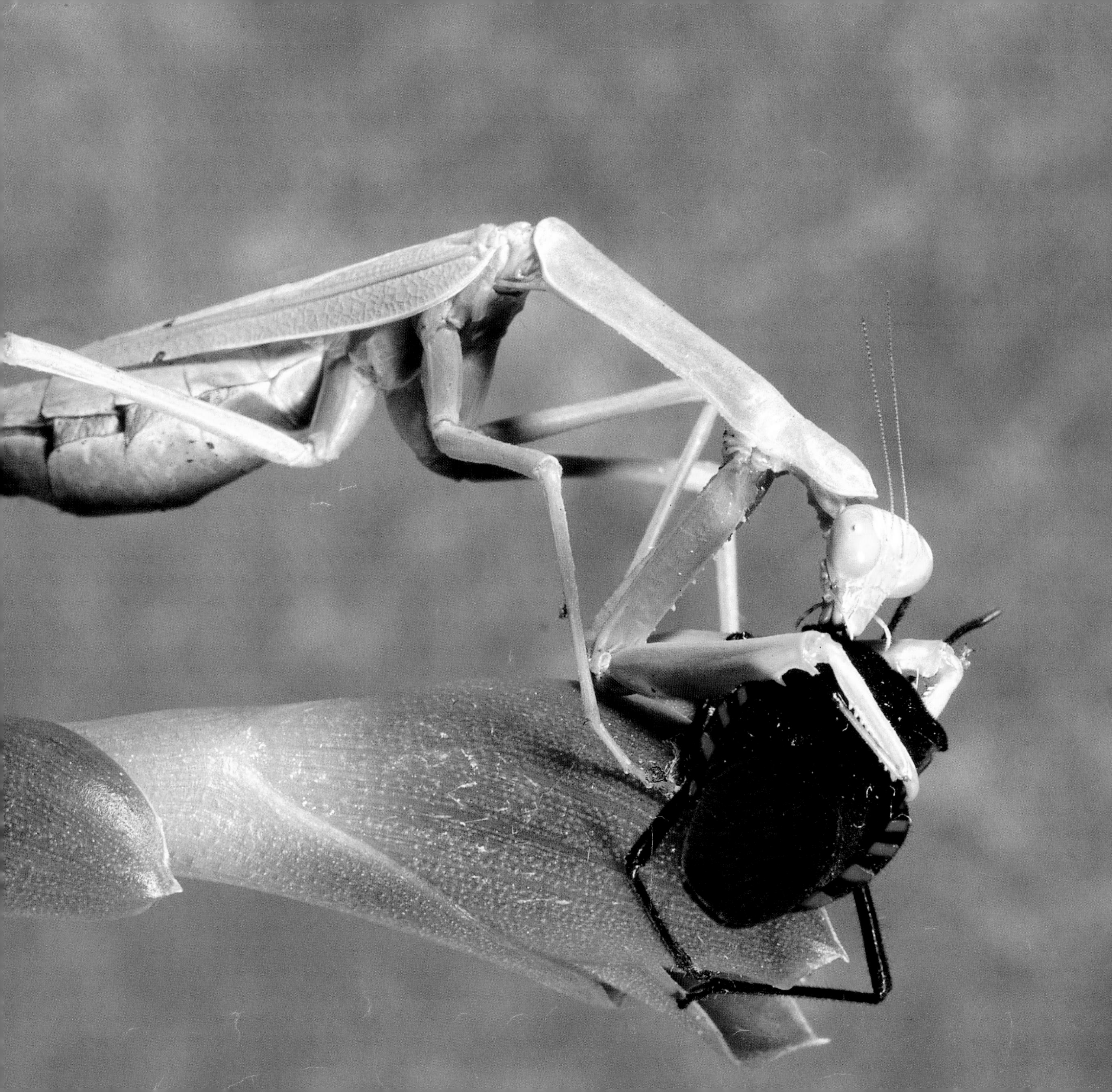

Praying mantises bite
and chew with strong jaws.

Las mantis religiosas
muerden y mastican con
unas fuertes mandíbulas.

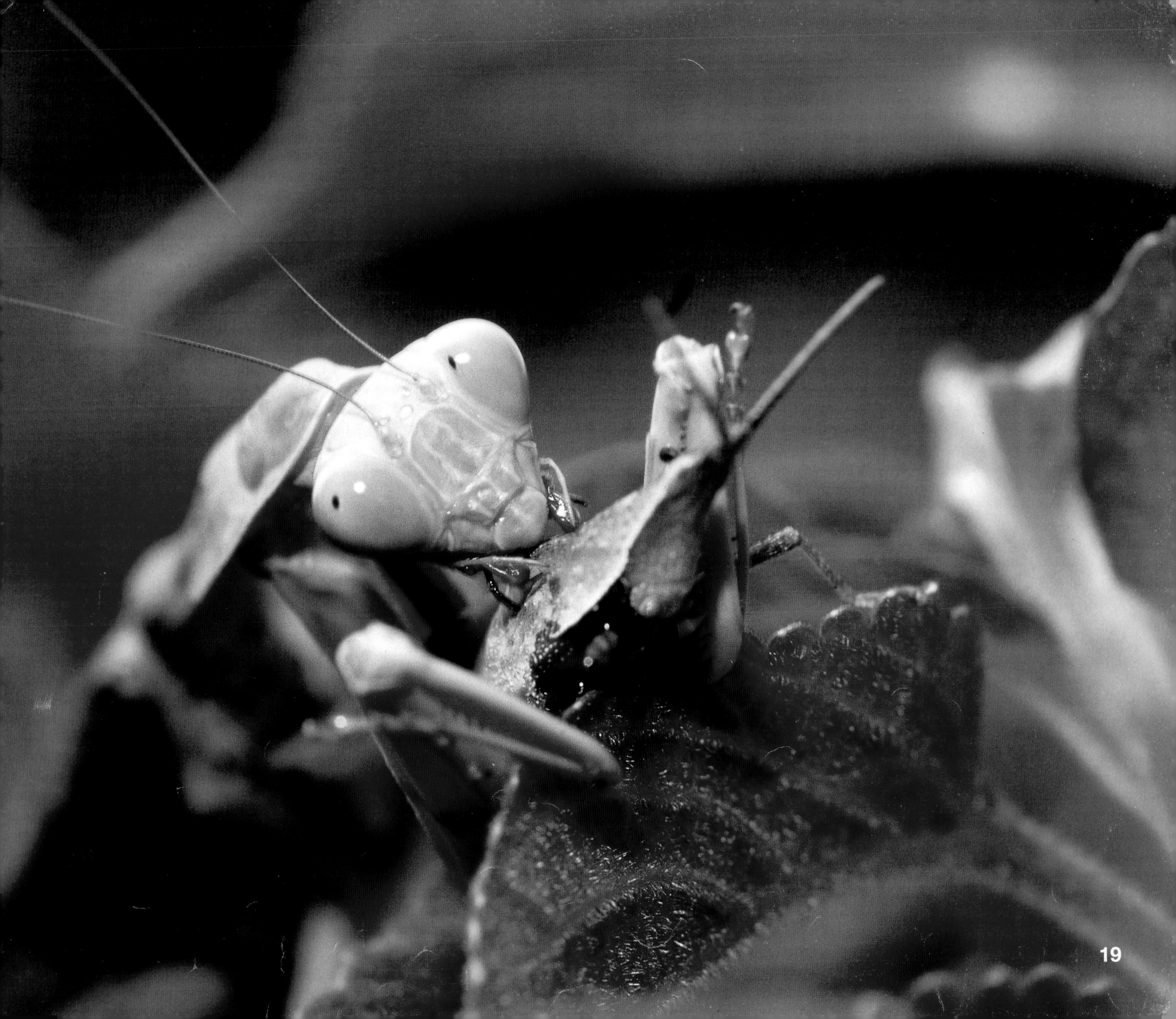

Praying mantises clean
themselves after they eat.

Las mantis religiosas se limpian
a sí mismas después de comer.

Glossary

fold—to bring together, or to bend close to the body; praying mantises fold their legs together; the action makes the mantis look like it is praying.

grab—to take hold of something quickly; praying mantises stay very still waiting for other animals to eat; they quickly grab the animals.

insect—a small animal with a hard outer shell, six legs, three body sections, and two antennas; most insects have wings.

jaw—a part of the mouth used to grab, bite, and chew

Glosario

atrapar—coger una cosa rápidamente; las mantis religiosas están muy quietas esperando comerse a otros animales.

doblar—apretar o juntar muy cerca del cuerpo; las mantis religiosas doblan y juntan sus patas; el movimiento hace que parezca que las mantis religiosas están rezando.

el insecto—animal pequeño con un caparazón duro, seis patas, cuerpo dividido en tres secciones y dos antenas; la mayoría de los insectos tiene alas.

la mandíbula—parte de la boca que se usa para atrapar, morder y masticar

Internet Sites

FactHound offers a safe, fun way to find Internet sites related to this book. All of the sites on FactHound have been researched by our staff.

Here's how:

1. Visit *www.facthound.com*
2. Choose your grade level.
3. Type in this book ID **0736866817** for age-appropriate sites. You may also browse subjects by clicking on letters, or by clicking on pictures and words.
4. Click on the **Fetch It** button.

FactHound will fetch the best sites for you!

Sitios de Internet

FactHound proporciona una manera divertida y segura de encontrar sitios de Internet relacionados con este libro. Nuestro personal ha investigado todos los sitios de FactHound. Es posible que los sitios no estén en español.

Se hace así:

1. Visita *www.facthound.com*
2. Elige tu grado escolar.
3. Introduce este código especial **0736866817** para ver sitios apropiados según tu edad, o usa una palabra relacionada con este libro para hacer una búsqueda general.
4. Haz clic en el botón **Fetch It**.

¡FactHound buscará los mejores sitios para ti!

Index

Índice